Royal Ruby

Philip Hopper

D1065844

WITHDRAWN

Schiffer Publishing Ltd

4880 Lower Valley Road, Atglen, PA 19310 USA

Dedication

I am dedicating this book to my mother and father, Philip and Margery Hopper. Throughout my childhood my mother always included me in her antiquing expeditions. While my mother provided me with the impetus to "collect," my father instilled a deep appreciation for the artistry and craftsmanship attributed to many antiques. Even today I am still awed by the intricate work crafted by the early artisans. I owe much of my success today to the love, devotion, and support my parents provided.

Designed by Bonnie M. Hensley
Typeset in Lydian Cursive BT/Korrina BT

ISBN: 0-7643-0667-7
Printed in China
1 2 3 4

Published by Schiffer Publishing Ltd.
4880 Lower Valley Road
Atglen, PA 19310
Phone: (610) 593-1777; Fax: (610) 593-2002
e-mail: schifferbk@aol.com

In Europe, Schiffer books are distributed by
Bushwood Books
6 Marksbury Avenue Kew Gardens
Surrey TW9 4JF England
Phone: 44 (0)181 392-8585; Fax: 44 (0)181 392-9876
e-mail: bushwd@aol.com

Please write for a free catalog.
This book may be purchased from the publisher.
Please include $3.95 for shipping. Please try your bookstore first. We are interested in hearing from authors with book ideas on related subjects.

Contents

Acknowledgments

This book would never have been published without the help of numerous glassware collectors and friends nationwide. I would like to make special tribute to several collectors. Rick Hirte (Sparkle Plenty Glassware, Bar Harbor, Maine) loaned me a rare 12 oz. High Point tumbler for the book. Francee Boches (Cheshire Cat Collectibles, Miami, Florida) made a six-hour drive to northern Florida to purchase the High Point pitcher for me, the only High Point pitcher I have seen in twenty-five years. Lyla E. Dyer (Grapevine Antiques, Lusby, Maryland) located many unique regional pieces of Royal Ruby glassware for the book. Finally, Barbara W. Birge (Lexington, Kentucky) not only located glassware for the book, but her e-mail messages to me kept my spirits up and my mind focused on completing this project. Without their help, the book would not have become a reality.

I would also like to thank many of the present and former employees of Anchor Hocking for helping me identify and document information about Royal Ruby glassware. I want to express my sincere appreciation to Phillip Bee for providing some of the production dates, David A. Bates for providing insight on glass batch formulation and the "striking" process, Mary Ann Jackson for providing many of the sales sheets for inclusion in the book, and Paul Stuart for identifying many of the unlisted pieces of Royal Ruby glassware.

Throughout the entire process of collecting the 1700 pieces of Royal Ruby glassware in my collection and writing this book, I was helped by two very close friends. They tolerated my collecting "obsession," constant "babbling" about glassware, and numerous visits to their home in Abilene, Texas, during glass buying trips. A special thanks to E.T. and Denise Moore.

Special appreciation is extended to Colonel Wiley Taylor for taking time out of his busy schedule to perform the preliminary review of the text.

Finally, I would like to thank the people at Schiffer Publishing Ltd. for making this project such an enjoyable and memorable experience.

Pricing

The prices in the book are only a guide. They are retail prices for mint condition glassware. Several factors will have an affect on glassware prices: regional availability, depth and consistency in Royal Ruby coloring, the presence or absence of Anchor Hocking markings in the glass or as paper labels, and relative rarity of the piece. Certain items will command higher prices if they are sets in the original packaging. Prices will drop considerably for glassware that is chipped, scratched, cracked, or deformed. No matter what any reference book states, the bottom line is...

Glassware is only worth what someone is willing to pay for it!

Measurements

I have tried to make this reference book as "user friendly" as possible. Too many times I have been in an antique shop and spotted a tumbler I wanted. The reference book I was using said this was a 12 oz. tumbler. Without a container of liquid and measuring cup I would have no way to actually determine if the tumbler held 12 ounces. I would rather know the tumbler is 5 inches high with a top diameter of 3 inches. This I can measure with a ruler. Unless otherwise noted, the measurements listed in the book are the height of the item. Realize that throughout the production of certain glassware items, the mold dimensions did vary. The measurements listed in the book are the actual measurements made on each piece of glassware pictured.

Regional Availability

Over the last seven years I have noticed regional differences in the availability of Royal Ruby glassware. Many items produced by Anchor Hocking were used as promotional items, and therefore were regionally distributed. These items were not listed in the catalog or "jobber" sheets used by sales personnel. For Example, the oval vegetable bowl is quite common in the midwest, but rare on the east coast. I can go into any antique mall and see stacks of the bowls. They are usually priced $37.50 and remain on the shelf for months. Another example, the "Howdy Wrangler" tumblers have only been located in the Oklahoma City, Oklahoma, or St. Louis, Missouri, areas. This local availability will definitely affect pricing!

Resources Available to Collectors

Collectors today have a great variety of resources available. With the advent of the "electronic age," collecting capabilities have been greatly expanded. I can honestly state this book would not have been possible without using the vast resources available, especially on the Internet. Below I have listed the resources collectors can use for locating antiques and glassware, however, realize this list is not all inclusive.

Internet Resources

Without leaving the comfort of your home or office, you can search worldwide for items to add to your collection. Presently, there are both antique dealers and auctions services on the Internet.

eBay Auction Service: The eBay Auction Service provides a continually changing source of items. This Internet service contains over 250,000 items in 371 categories. Internet users can register as both buyers and/or sellers. The majority of the items remain on the "auction block" for seven days. You can search the auction database for specific items. A list of items will be presented following the search. For example, you might want to find a Fire King Jadite vase made by Anchor Hocking. Because the seller enters the item's description in the database, you often have to anticipate how the item is described. Don't limit the searches. In this case, you might have to search under jadite, hocking, fireking (no space), fire king (with the space), or vase to find the item you want.

Internet Antique Malls: There are several Internet antique malls I have found to be extremely useful in locating glassware. Each mall contains numerous individual dealers with items for sale. The malls I used are listed below:

TIAS Mall – (http://www.tias.com/)

Collector Online Mall - (http://www.collectoronline.com/)

Facets Mall - (http://www.facets.net/facets/shopindx.htm)

Depression Era Glass and China Megashow – (http://www.glassshow.com/)

Cyberattic Antiques and Collectibles – (http://cyberattic.com/)

Glass Shows, Antique Shops, and Flea Markets

All collectors still enjoy searching the deep dark crevices of the local antique shops and flea markets. Many of the best "finds" in my personal collection were located in flea markets and "junk" shops. I recently found a crystal Anchor Hocking High Point pitcher and tumblers in a flea market. The High Point pattern is rare in Royal Ruby and was never marketed in crystal. The crystal pitcher and glasses were probably "test" pieces made prior to the introduction of the Royal Ruby High Point. Most of the dealers in glass shows have a good working knowledge of glassware, so "real finds" are not too plentiful.

Periodicals

Both the *Depression Glass Magazine* and *The Daze, Inc.* are periodicals which will greatly enhance your collecting abilities. Along with the numerous advertisements for glassware, there are informative articles on all facets of collecting glassware.

Word of Mouth

This is one resource so often overlooked. Let others know what you are looking for. Consider expanding your search by including friends, relatives, and other collectors. This book could not have been written without the help of many fellow collectors.

Do not limit your collecting to only one resource. Remember the items you seek are out there... somewhere!

Request for Additional Information

I am always seeking information concerning Anchor Hocking's Royal Ruby glassware. Much of the information about Royal Ruby is not available outside the company. This book will undoubtedly be updated, and it is imperative new information be made available to collectors. Feel free to contact me at the following address:

Philip L. Hopper
1120 Choctaw Ridge Road
Midwest City, OK 73130
E-mail: rrglass@swbell.net

Please be patient if you need a response. I am not in the glassware business. I am a military officer first and a collector the rest of the time. I will make every effort to provide prompt feedback on your inquiries.

Crystal and Green Glassware

Several pieces of non-Royal Ruby glass have been included only for pattern identification or illustration. Variations in the "cubes" on Windsor tumblers or the pattern on the High Point tumblers are hard to see because of the dark red color. The Old Café relish trays/lazy susans were photographed with some crystal inserts to show the detail in the tray. While buying glassware, I found a crystal Old Café insert with no lines on the bottom. I couldn't figure out whether it was a "mistake" made during production or an insert that fit a specific relish tray. When I found the ornate silver plated relish set in Lubbock, Texas, it contained the inserts without the bottom lines. They were made so the buyer could see the extremely ornate engraving in the silver tray. This engraving is not visible with the lined inserts. Again, this is my attempt to make this book "user friendly."

Producing the Royal Ruby Color

Many collectors believe all Royal Ruby glass produced by Anchor Hocking was made using gold to give the glass the familiar deep red color. Many of the pieces made prior to 1950 did contain gold, however, Royal Ruby glass made after that date was made with a glass "batch" mixture containing bismuth, tin, and copper. As the "batch formula" was refined, bismuth was eliminated with no effect on the color. When the Royal Ruby glass was first molded or pressed into shape, it had a very light green color. Once the glass was removed from the molds it was transferred to an annealing oven, called a lehr. The glass was placed in a lehr and the temperature curve adjusted to re-heat the glass to 1100 degrees Fahrenheit for 15-20 minutes. The temperature was reduced gradually over the next 1.5 to 2 hours to anneal, or reduce, the internal stresses in the glass. The change from light green to deep red color, termed "striking," occurred during the first 15-20 minutes in the lehr. Urea, added to the original glass batch, acted as a reducing agent and changed the glass from a light green to a deep red color.

Rainflower vase after annealing.

Most collectors of Royal Ruby glassware have noticed extreme color variations. Some pieces actually have clear areas where the reduction process failed to occur. Three factors effected the "striking" process: temperature, time in the lehr, and the amount of urea in the batch. If the urea level was too low, the glass appeared too light or would not strike. High urea levels caused the ruby color to be too dark. Early lehrs were equipped with asbestos curtains to control the temperature in the striking zone of the annealing process. On some glasses, the thicker bottom areas and rims actually "struck" during the molding process. The Rainflower vase pictured to the left was partially "struck" when molded. Only the thicker portions remained hot enough before annealing to turn red. If this vase had been placed in the lehr and allowed to reach 1100 degrees Fahrenheit for 15-20 minutes, the entire piece would "strike" and would look like the vase above on the right.

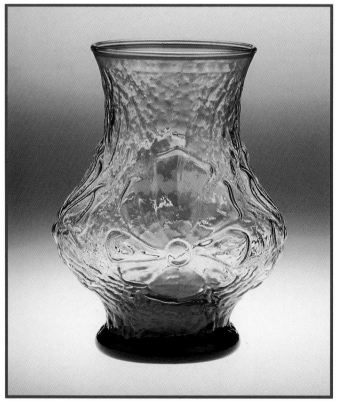

Rainflower vase after removal from the mold.

Identification Marks

Over the years Anchor Hocking has used several identification marks to mark their glassware. In 1980, the company issued a limited edition 75th anniversary ashtray, pictured below, which portrays the corporate identification marks. The marks on the ashtray were blackened with a magic marker so they would show up when photographed. Originally, when the Hocking Glass Company was established in 1905, the company used the mark seen on the left side of the ashtray. This mark was used from 1905 until 1937, when it was replaced by the more familiar "anchor over H" mark (center of ashtray) to illustrate the merger of the Hocking Glass Company and the Anchor Cap Company. Finally, in October 1977, the company adopted a new symbol (right side of the ashtray), an anchor with a modern, contemporary appearance to further the new corporate identity.

Not all Royal Ruby glassware was identified by marks incorporated directly into the glass. At least three styles of paper labels were attached to glassware items.

Crystal ashtray, 5.5" in diameter, $50-55 with original box

Paper label.

75th Anniversary ashtray.

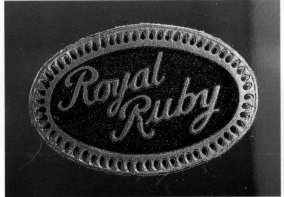

Paper label.

Paper label.

Chapter One

Established Patterns

Baltic

The commonly found items are the "footed" glasses and sherbets. The 12 oz. footed bowls, while not as common, were listed in the 1971 Anchor Hocking catalog, but shown in crystal only.

Listing for the Baltic Pattern in the 1971 Anchor Hocking catalog.

Baltic tumblers. Left to right: 10 oz. goblet #R3316, 4 1/2", $8-10; 5 oz. juice #R3311, 3 5/8", $5-8; 3 1/2 oz. cocktail #R3312, 2 5/8", $5-10.

The Baltic footed bowl may be confused with the popcorn bowl. The Baltic 12 oz. footed bowl #R3322 on the left has a diameter of 4 3/4" and much thinner glass, $10-15. The popcorn bowl on the right has a diameter of 5 1/4" and the glass is considerably thicker, $15-20. Also note the bases of the bowls are different.

Here are the three sherbets produced by Anchor Hocking. On the far right is the Baltic 6 1/2 oz. sherbet #R3313. It was originally sold with a 6 1/4" plate #R828. This sherbet is 2 3/8" tall and 3 1/2" in diameter, $5-10.

Berwick

Berwick was made from the 1950s to 1970s. This pattern was listed in the 1971 Anchor Hocking catalog, but only in crystal. Some collectors call this pattern either "crystal foot" or "boopie." The shape of the bowl, stem, and base differs from the Early American stemware. Quite common four or five years ago, Berwick stemware has virtually disappeared because it makes a popular table setting to use during the Christmas holiday season.

Berwick stemware. Left to right: 9 oz. goblet #R736, 5 1/2", $15-20; 4 oz. juice/wine #R735, 4 1/2", $12-15; 6 oz. sherbet #R733, 3 1/2", $8-12; 3 1/2 oz. cocktail #R734, 3 7/8", $12-15.

Anchor Hocking produced two types of crystal stemware. There are several differences between the stem of the Early American (bubble) on the left and Berwick (boopie) on the right. The shape of the stem and the number and type of "bumps" on the two bases vary considerably.

Beverly

This pattern was made in six sizes of glasses. The catalog lists the glasses in crystal and I have only been able to find nine of the 4 1/2 oz. cocktails in Royal Ruby.

Anchor Hocking listed six pieces of Beverly in the 1971 catalog. I have only been able to find the 4 1/2 oz. cocktail in Royal Ruby.

Beverly 4 1/2 oz. cocktail in Royal Ruby, #R3264, 2 3/4", $10-12.

Burple

Anchor Hocking only made the 4 1/2" and 8" bowls in the Burple pattern. They were sold as the Burple Dessert Set which included one 8" and six 4 1/2" bowls. The bowls are common in crystal and green, but relatively rare in Royal Ruby.

Burple glassware has a very distinctive pattern. Here is the 8" berry bowl #R1878, $20-30, and the 4 1/2" dessert bowl or nappy, #R1874, $8-15.

Anchor Hocking sold this pattern as the Burple Dessert Set consisting of one 8" and six 4 1/2" bowls, $70-75.

Charm

Charm (square) dinnerware was produced in the 1950s and given the R-2200 designation by Anchor Hocking. There are several references to the 9 1/4" dinner plate, but I have been unable to confirm its existence. Also, I have never seen either the 6" soup bowl or the 8" x 11" platter in Royal Ruby. I seriously doubt any of these exist, however, there is always the possibility some "test" pieces were made.

Cup #R2279, $4-5; saucer #R2229, $4-5.

Luncheon plate #R2241, 8 3/8", $10-15

Left to right: dessert bowl #R2275, 4 3/4", $8-10; salad bowl #R2277, 7 1/2", $15-20. I also have both sizes of these bowls with etched tulips on two sides of each bowl.

Classic

Classic was made in the 1960s. The pattern, often called "Rachael" by collectors, was given the R-900 designation by Anchor Hocking. Only the 10" vase, 11" round bowl, 14 1/2" under liner plate, and 8" x 12 1/4" oval bowl have been found in Royal Ruby Classic.

Oval bowl, 8" by 12 1/4", $30-40.

Round bowl, 11", $40-50; under liner plate, 14 1/2", $50-65.

Classic vase, 10", $40-50.

Coronation

Coronation was made from 1936 to 1940. Collectors often refer to this pattern as "Banded Rib" or "Saxon." Coronation bowls are the most common Royal Ruby items available. The three sizes of bowls, the 8", 6 1/2", and 4 1/2", have open handles. Usually the bowls were sold as a 7-Piece Dessert Set #A4400/8. The set included one 8" and six 4 1/2" bowls. The Royal Ruby coffee cup was sold with the crystal saucer/sherbet plates. To date, no Royal Ruby saucers/sherbert plates have been found. The crystal saucers/sherbet plates are the only pieces of crystal Coronation found.

The Royal Ruby Coronation coffee cups were sold on 6" crystal saucers/sherbet plates, $4-5. During the preparation of this book, the only Royal Ruby Coronation coffee cup I have could not be located.

Berry bowl, 8", $10-15; nappy, 6 1/2", $12-15; dessert bowl or nappy, 4 1/2", $5-8.

The Coronation bowls were sold in a 7-Piece Dessert Set #A4400/8. The set consisted of one 8" and six 4 1/2" bowls, $40-50.

Early American

Although Anchor Hocking gave this stemware the name Early American, collectors still use the name "bubble." Early American was made in a 13 oz. ice tea, 9 1/2 oz. goblet, 6 oz. sherbet, 5 1/2 oz. juice, and 4 1/2 oz. size cocktail. Crystal and Forrest Green pieces were made in all five sizes, but Royal Ruby was only made in four sizes. The 13 oz. ice tea was not produced in Royal Ruby. The shape of the bowl, stem, and base differs from the Berwick stemware. Quite common four or five years ago, Early American stemware has virtually disappeared because it makes a popular table setting during the Christmas holiday season.

Anchor Hocking produced two types of crystal stemware. There are several differences between the stem of the Early American (bubble) on the left and Berwick (boopie) on the right. The shape of the stem and the number and type of "bumps" on the two bases vary considerably.

Left to right: 10 oz. goblet #R336, 5 1/4", $15-20; 4 1/2 oz. fruit juice #R335, 4 1/4", $12-15; 6 oz. sherbet #R333, 4", $12-15; 3 1/2 oz. cocktail #R334, 3 3/8", $18-20.

Early American Prescut

Even though Anchor Hocking made countless Early American Prescut pieces from 1960 to 1978, the company only made one piece in Royal Ruby, the 7 3/4" ash tray. The Royal Ruby ash tray is trademarked in the glass.

Ashtray, 7 3/4", $20-30.

Fairfield

Anchor Hocking produced a variety of Fairfield colors through the years: avocado, honey gold, laser blue, sky blue, crystal, spearmint, smoke gray, spicy brown, and Royal Ruby. Royal Ruby pieces, given the R-1200 designation, appear to be one of the harder colors to find. Unlike all the other colors that were not trademarked, Royal Ruby pieces were trademarked.

The trademark, the familiar "anchor over H" emblem, is located on the side of the base for the relish dish and 5 1/4" two-handled bowl. On the compote, the trademark is located in the center of the base. There is a depression going up into the stem and the emblem is located in the depression.

Compote, 3 3/4", $12-15.

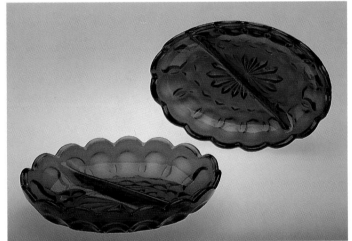

Divided relish dish, 5" x 7", $12-15.

Two handled bowl, 5 1/4", $15-20.

Fortune

The Fortune pattern was produced from 1937 to 1938. While the majority of the pieces were produced in crystal and pink, the candy dish was made with a Royal Ruby cover. The Royal Ruby cover does not have the paneling normally found on the candy dish cover. Also notice the patterns in the bottom of the crystal and pink candy dishes are different.

Pink candy dish with Royal Ruby cover, 5 1/2", $15-20. Notice the pattern in the bottom of the pink dish is different than the crystal dish.

Crystal candy dish with Royal Ruby cover, 5 1/2", $15-20. The Royal Ruby cover is not paneled like most commonly found Fortune candy dish covers. Note the quilted pattern in the bottom of the crystal dish.

Crystal candy dish with Royal Ruby lid in place, $15-20.

Georgian

The Georgian pattern was probably produced from the 1940s to 1970s, but not continuously throughout this period. Unlike many of the other patterns of Royal Ruby, some of the tumblers have the "anchor over H" emblem embedded in the bottom of the glass. The tumblers tend to chip very easily, therefore, mint pieces will command a premium price. This is the only Royal Ruby pattern where Anchor Hocking made salt and pepper shakers.

Left to right: Georgian pitcher, 80 oz., $40-50; 10 oz. pedestal tumbler, 5 1/2", $10-15; 14 oz. tumbler, 5 1/2", $10-15; 12 oz. tumbler, 5", $10-15; 9 oz. tumbler, 4 1/4", $10-15; 5 oz. tumbler, 3", $8-10; 5 oz. tumbler, 3 1/4", $8-10. The smallest two tumblers each hold 5 ounces. This is possible because the shorter tumbler has a diameter 1/4" greater than the taller tumbler.

Box containing 1 dozen 9 oz. tumblers, $15-20 for box only.

Salt and pepper shakers #R45,
4", $40-50 a pair.

Box which contained two dozen salt and pepper shakers, $50 for box only.

High Point

High Point was another pattern made early in Anchor Hocking's history of Royal Ruby glass production. Produced in the early 1940s, it had similar problems with very light red or clear areas in the glass. Most of the pieces in this pattern are uncommon. The 5 oz. and 9 oz. tumblers can be found, but the 13 oz. tumbler is rare. Until recently, I thought High Point was only made in Royal Ruby. I have found the 5 oz. fruit juice, 9 oz. table tumbler, and pitcher in crystal. These were probably test pieces produced before the line was introduced to the public.

Left to right: water pitcher #R1287, 2 1/2 qt., $75-100; 13 oz. ice tea #R1208, 5 1/2", $35-40; 9 oz. table tumbler #R1201, 4 1/4", $12-15; 5 oz. fruit juice #R1203, 3 5/8", $10-15. *13 oz. ice tea #R1208 courtesy of Rick Hirte.*

High Point was also made in crystal. I have found the pitcher (not shown) and the 5 oz. and 9 oz. tumblers pictured here.

Tumbler grouping which shows the High Point pattern often hard to see in most photographs due to the deep red color of the glass.

Hobnail

This was Anchor Hocking's first attempt to produce Royal Ruby glassware. Produced in the late 1930s, many of the pieces had very light red or clear areas. As I explained earlier in the book, the process of "striking" the ruby red color is extremely time, temperature, and batch formula dependent. It is evident the process had not been perfected when many of the Hobnail pitchers and tumblers were made.

Company advertising sheet introducing the Hobnail pattern.

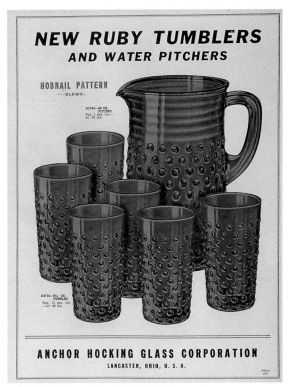

Left to right: pitcher #A2744, 60 oz., $30-35; 9 1/2 oz. tumbler #A2719, $8-10.

Manhattan

Anchor Hocking produced Manhattan, also known as "horizontal ribbed," from 1938 to the mid 1940s. Crystal and pink are the commonly found colors. Only the inserts for the relish trays are common in Royal Ruby. The inserts were also used with the 14" crystal Waterford plate.

Manhattan crystal relish tray with Royal Ruby inserts, complete set, $50-65. Ruby inserts separately, $5-8; 14" crystal plate, $10-15. The inserts were also used with the Waterford pattern listed later in this chapter.

Old Café

Anchor Hocking produced Old Café glassware from 1936 to 1940. Like Coronation, the Royal Ruby coffee cup was sold with a crystal saucer. The 8" footed tray #A977, often called the low candy dish, was sold with a metal handle. The Old Café lamp and vase, not pictured in this book, are reasonable rare. I have recently located the Old Café lamp, but not in time to be included in the book. It will be included in the next edition of this book.

Berry bowl, 3 3/4", $15-18.

Sherbet, 3/4" low footed, $12-15.

Cereal bowl, 5 1/2", $12-15.

Old Café bowls pictured together so you can see the relative size of each bowl.

Left to right: water tumbler, 4", $25-35; juice tumbler, 3", $20-25.

Royal Ruby cup, $4-6; 6" crystal saucer, $2-4.

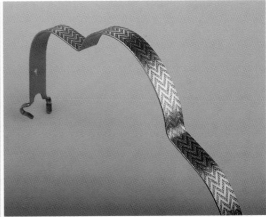

Closeup of the footed tray #A977 handle so you can see the pattern stamped in the metal.

Footed tray #A977, 8", $12-15 with metal handle, $8-10 without metal handle. The metal handles were produced by the Heller Company of Brooklyn, New York.

The complete relish set, $125-150, consists of one 15" crystal plate, $30-35, five inserts, $8-10, one 7" covered dish, $10-15, and one 4" two-piece rotating base, $30-40.

Crystal plate, 15", $30-35. There is a ridge around the clear area in the center of the plate. This ridge holds the base in place so the tray doesn't move off center and tip over.

Crystal two-piece rotating base for the relish set, $30-40.

Oyster and Pearl

Oyster and Pearl was produced from 1938 to 1940. The pattern was produced in pink, crystal, Royal Ruby, and white fired on either green or pink. Only six items were produced in Royal Ruby. The 5 1/4" heart-shaped, 1-handled bowl and 5 1/2" round, 1-handled bowl are often confused. The heart shaped bowl has a small pour spout on the side opposite the handle. Also, the area where the handle is attached is slightly flattened, giving rise to the "heart shape." I wanted to include a side-by-side photo of the two bowls, however I could not locate the 1-handled, heart shaped bowl in Royal Ruby.

Oyster and Pearl 1-handled bowl, 5", $10-15.

Oyster and Pearl nut dish #A875, 5 1/4", $12-15.

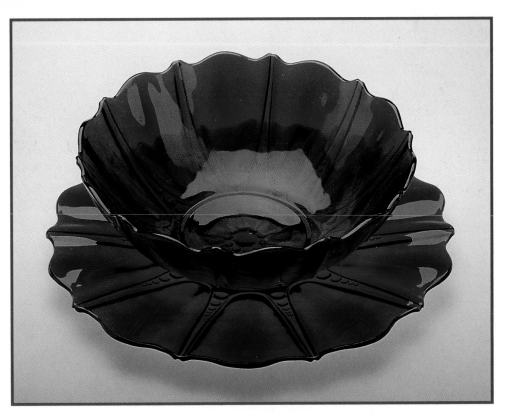

Console bowl #A889, 10 1/2", $30-40; sandwich plate #A890, 13 1/2", $30-40.

Candle holder #A881, 3 1/2", $50-60 per pair.

Console set, $125-150.

Provincial

Royal Ruby Provincial, commonly known as "bubble" by collectors, was designated as the R-1600 series that was produced in 1962 to 1963. The pitcher and glasses were sold in a variety of sets. They were sold in the R1600/55 (9-Piece Refreshment Set), R1600/56 (24-Piece Hostess Service Set), R1600/57 (25-Piece Refreshment Set), R1600/58 (24-Piece Hostess Service Set, and a 15-Piece Refreshment Set. So far, no Royal Ruby sugars or creamers have been found. Most of the Provincial pieces are marked with paper labels, however, I have found the 12 oz. tumbler #R1612 and 6 oz. fruit juice #R1606 with the "anchor over H" emblem embedded in the bottom of the glass.

Dinner plate #R1641, 9 1/4", $20-25.

Left to right: ice lip pitcher #R1660, 64 oz., $50-60; 16 oz. ice tea #R1616, 5 3/4", $18-20; 12 oz. tumbler #R1612, 4 1/2", $12-15; 9 oz. old fashioned #R1609, 3 1/4", $12-15; 6 oz. fruit juice #R1606, 3 3/4", $8-12.

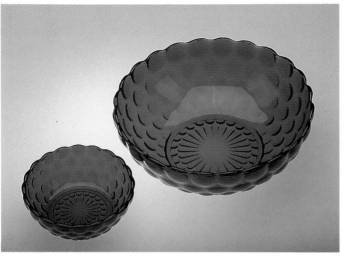

Left to right: dessert bowl #R1664, 4 1/2", $8-10; berry bowl, 8", $15-20.

Cup #R1650, $5-10; saucer #R1628, $5-10.

2-tier tidbit, $50-60.

9 Piece Refreshment Set; $20-30 for box only.

9 Piece Refreshment Set #R1600/55, $170-185 with box.

Queen Mary

The Queen Mary pattern, produced from 1936 to 1949, is usually found in crystal and pink. The pattern is also called "vertical ribbed" by collectors. The 3 1/2" round ash tray was made in Royal Ruby in the 1950s and is common today. The 4 1/2" double branched candlesticks are hard to find and you can expect to pay considerably more for the Royal Ruby candlesticks than you will for the crystal candlesticks.

Ash tray, 3 1/4", $5-8.

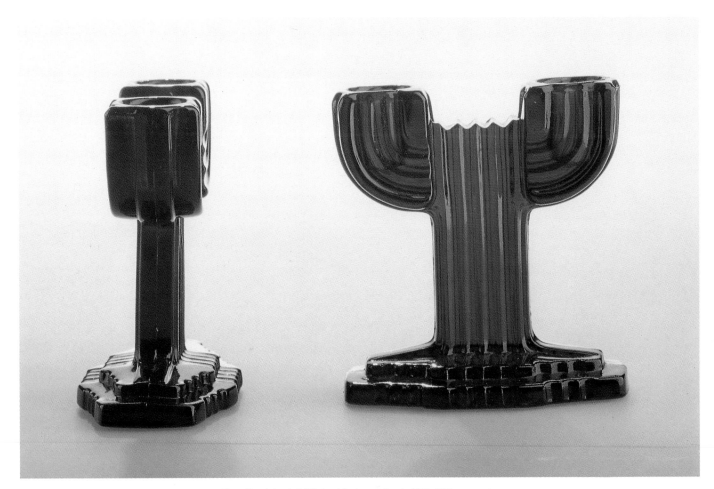

Candlestick, 4 1/2" double branched, $75-125 per pair.

Rainflower

Rainflower was first introduced in 1973 and continued until 1978. Only the 5 1/2" vase was produced in Royal Ruby and all vases I have found are trademarked. Because the bottom of the vase has many imperfections, the extremely small trademark is difficult to see. Using a magnifying glass, the familiar "anchor over H" emblem can be found.

Vase, 5 1/2", $10-15. All the vases are marked with an extremely small "anchor over H" emblem embedded in the glass. Because the bottom of the vase has many imperfections, the mark is hard to see without a magnifying glass, but it is there.

R-1700

The R-1700 pattern was produced in the 1940s. Most of the pieces were marked with paper labels.

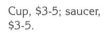

Cup, $3-5; saucer, $3-5.

Left to right: salad plate, 7 3/4", $5-8; dinner plate, 9 1/8", $8-10.

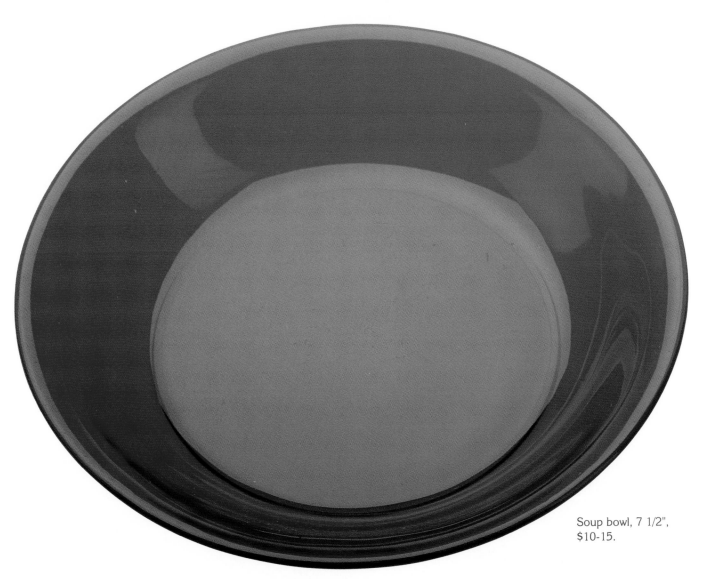

Soup bowl, 7 1/2",
$10-15.

Left to right: sugar without lid, $8-10; creamer, $8-10.

R-4000

The R-4000 was not listed in any catalog. It was probably offered as a special order item. The only boxed set I have ever seen is the 22-Piece Luncheon Set pictured. Some of the pieces, namely the 6 5/8" cereal bowls, are scarce.

Cup, $3-5; saucer, $3-5.

Left to right: salad plate, 7 1/4", $8-10; dinner plate, 9", $12-15.

Cereal bowl, 6 5/8", $15-20.

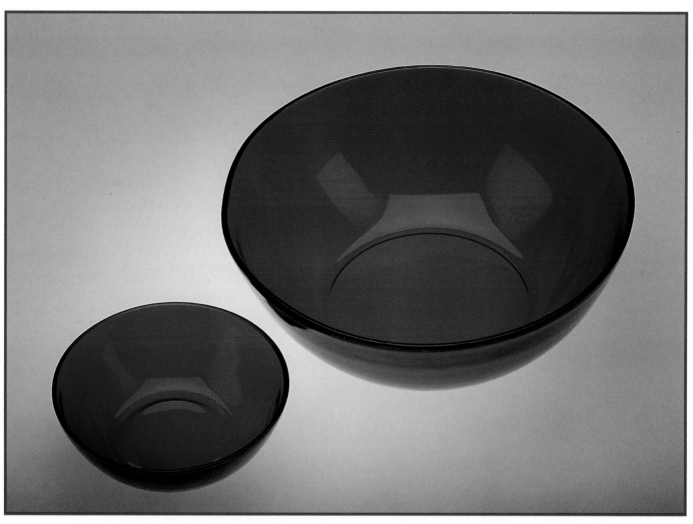

Left to right: dessert bowl, 4 1/2", $5-6; vegetable bowl, 8 1/4", $20-25.

Left to right: sugar, $8-10; creamer, $8-10.

22-Piece Luncheon Set, $350-375 for the complete set in the box; $25-30 for box only.

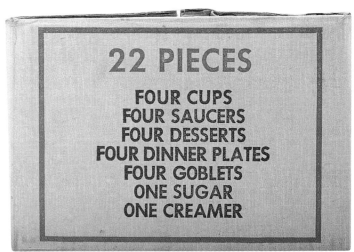

Side view of luncheon set listing contents.

Contents of the 22-Piece Luncheon Set plus the salad plate.

Roly Poly

This pattern was undoubtedly made for several years. Given the R3600 designation, Roly Poly was sold in at least three different sets: 9-Piece Iced Tea Set #R3600/2, 24-Piece Refreshment Set #R3600/4, and the 19-Piece Refreshment Set #R3600/5. The 1971 catalog lists two other sizes of Roly Poly, the 9 oz. on-the-rocks and the 6 oz. juice. These are listed in crystal in the catalog but they may exist in Royal Ruby.

Left to right: 96 oz. upright pitcher, #R3687, $35-45; 13 oz. beverage/ice tea #R3658, 5", $8-10; 11 oz. tumbler, 4 3/4", $8-10; 9 oz. table tumbler #R3651, 4 1/4", $8-10; 5 oz. fruit juice #R3653, 3 3/8", $5-8.

9 Piece Refreshment Set, $100-125 for complete set in the box; $20-25 for box only.

Refreshment set removed from the box.

Juice set consisting of the 42 oz. tilted pitcher and six 5 oz. juice glasses, $75-80.

There are two versions of each of the three different tilted pitchers. Here are the two versions of the 42 oz. tilted pitcher.

Closeup of the 42 oz. pitchers. The earlier version on the left has vertical lines on the neck and a smooth handle. The later version on the right has horizontal lines going around the neck and the handle is ribbed.

Sandwich

Royal Ruby Sandwich glassware was produced in the late 1930s. Only the 8 1/4" scalloped bowl, 6 1/2" scalloped bowl, 5 1/4" scalloped bowl, and the 5 1/4" smooth bowl were produced in Royal Ruby. I was not able to locate the 6 1/2" scalloped bowl for the book. Hopefully, it will be included in the first revision.

Scalloped bowl, 8 1/4", $35-45.

Scalloped bowl, 5 1/4", $18-20.

Smooth bowl, 5 1/4", $18-20.

Three of the four Sandwich bowls together. The two smaller bowls have the same diameter and line pattern around the base and rim. The scalloped bowl is 1/8" taller than the plain bowl.

Shell

Anchor Hocking produced Shell glassware in crystal from 1976 to 1978. During this period, at least one size of plate was produced in Royal Ruby, the 6" salad plate. I have thirteen of these plates and they are all trademarked with the familiar "anchor over H" emblem embedded in the glass. The emblem is located on the upper surface near the handle. Other pieces of Royal Ruby Shell may have been made.

Salad plate, 6", $12-15. Unlike crystal Shell glassware, the Royal Ruby pieces are marked. I have thirteen plates and each has the "anchor over H" emblem on the top surface just inside the rim near the handle.

Swirl

Only the pitcher and 10 oz. tumblers were made in Royal Ruby. Over the years, two different versions of the 80 oz. jug (pitcher) were produced. The earlier version of the pitcher has vertical lines on the neck and a smooth handle. This version was pictured in the 1954 catalog as the 80 oz. jug #A1087. The later version has horizontal lines going around the neck and the handle is ribbed.

Left to right: 80 oz. jug #A1087, $35-45; 10 oz. tumbler #A1031, $8-10.

The Swirl 80 oz. jug #A1087 was produced in two versions. Anchor Hocking actually listed the item as a "jug" and not a pitcher in the catalog.

Closeup of the Swirl 80 oz. jugs. The earlier version on the left has vertical lines on the neck and a smooth handle. The later version on the right has horizontal lines going around the neck and the handle is ribbed.

Waterford

This pattern was not made in Royal Ruby, however, the Royal Ruby Manhattan relish tray inserts were used with the 14" crystal plate.

Manhattan Royal Ruby relish tray inserts were also used with the Waterford crystal 14" plate, $50-60.

Whirly Twirly

Whirly Twirly glassware was produced in both Forrest Green and Royal Ruby. The Royal Ruby glassware is relatively hard to find, while the Forrest Green is reasonably common. Whirly Twirly (notice it is not spelled Whirley Twirley) glasses are often confused with the 9 1/2 oz. tall tumbler #R3597 and smaller juice glass. I have included side-by-side photographs to show the differences.

Left to right: 3 quart water pitcher, $75-80; 12 oz. tumbler, 5", $15-20; 5 oz. tumbler, 3 1/2", $12-15.

Left to right: 12 oz. tumbler, 5", $15-20; 9 oz. tumbler, 4", $15-20; 5 oz. tumbler, 3 1/2", $12-15.

Two tumblers often confused. On the left is the Whirly Twirly 12 oz. tumbler (5" tall) and on the right is the 9 1/2 oz. tall tumbler #R3597 (4 3/4" tall). Notice the Whirly Twirly tumbler's sides are tapered and the base is footed. The R3597 tumbler has straight sides and no foot on the base.

Dessert bowl, 3 3/4", $8-12.

Windsor

Anchor Hocking produced Windsor glassware in the 1940s. The 60 oz. pitcher #A1153, 9 oz. tumbler #A1131, and 5 oz. fruit juice #A1133 are generally found in Royal Ruby. There may be variations in the number of rows of "cubes" on Royal Ruby glasses, since I have found at least three different variations in crystal and light green. Also, I have purchased a light green 4 3/4" 12 oz. tumbler with four rows of "cubes." This is a previously unknown size which may have been produced in Royal Ruby.

Factory sheets announcing the Windsor pattern.

Left to right: 60 oz. pitcher #A1153, $35-40; 9 oz. tumbler #A1131, 4", $8-10; 5 oz. fruit juice #A1133, 3 1/4", $8-10.

Closeup of the crystal and Royal Ruby 9 oz. tumblers showing the "cube" pattern on the base of the glass.

Left to right: three versions of the 9 oz. tumbler. Notice there are three different styles of "cubes" on the base: 2, 3, and 4 rows. These versions may also exist in Royal Ruby.

Chapter Two
Bottles

Water Bottles

Three types of bottles are listed in this chapter: water bottles, beer bottles, and medicine bottles. The water bottles are common in crystal, reasonably hard to find in Forrest Green, and rare in Royal Ruby. The Royal Ruby beer bottles were manufactured for Schlitz. The bottles were made in 1949, 1950, and 1963. The dates of manufacturing are listed on the bottom of the bottle as a two-digit number. Also, there is the familiar "anchor over H" emblem, the words "Royal Ruby Anchorglass," and the number 5 to indicate the place of manufacturing (plant 5 in Connelsville, Pennsylvania). I also included trial bottles never marketed and two types of medicine bottles. These bottles are an interesting part of Anchor Hocking's production of Royal Ruby glassware.

Ribbed water bottle with lid, $200-250.

Plain water bottle with lid, $200-250.

Beer Bottles

Right: 32 oz. Schlitz beer bottle, 9 3/4", $50-65. The bottle is labeled "NOT TO BE RE-FILLED" and "NO DEPOSIT – NO RETURN" around the outside of the base. On the bottom of the bottle is the mold design (8585C), 5 (place of manufacturing - plant 5), Anchor Hocking emblem (anchor over H), 49 (year of manu-facturing—may also be 50 or 63), and the saying "Royal Ruby Anchorglass."

All ten beer bottles lined up so you can see the relative size of each. Left to right: 32 oz. Schlitz; 16 oz. Schlitz; 12 oz. Schlitz; 7 oz. Schlitz; two versions of foreign export beer bottle; Pfeiffer's Glasolite; three experimental beer bottles.

Closeup of the 32 oz. Schlitz beer bottle taken so you can see the shape of the mouth of the bottle.

12 oz. Schlitz beer bottle, 5 5/8", $30-35. The bottom of the bottle is marked with the mold number (168-38B), 5 (place of manufacturing—plant 5), Anchor Hocking emblem (anchor over H), 63 (year of manufacturing—may also be 49 or 50), and the saying "Royal Ruby."

16 oz. Schlitz beer bottle, 7 1/2", $100-110. The neck of the bottle is labeled "NOT TO BE REFILLED" and "NO DEPOSIT*NO RETURN." On the bottom of the bottle is the mold design (168-50), 5 (place of manufacturing—plant 5), Anchor Hocking emblem (anchor over H), 63 (year of manufacturing—may also be 49 or 50), and the saying "Royal Ruby."

The 12 oz. and 16 oz. Schlitz beer bottles have the same shaped mouth.

Closeup of the mouth of the 7 oz. Schlitz beer bottle.

7 oz. Schlitz beer bottle, 8", $30-40. The bottom of the bottle is marked with the mold number (67-22), 5 (place of manufacturing—plant 5), Anchor Hocking emblem (anchor over H), 50 (year of manufacturing—may also be 49 or 63), and the saying "Royal Ruby Anchorglass."

Closeup of the mouth of the foreign export bottle. On this version, the quilting on the bottle does not extend into the area of the neck where the printing is located. On another version, the printed area in the neck is covered with quilting.

12 oz. Beer bottle, 8", $50-55. The neck of the bottle is labeled "NOT TO BE REFILLED" and "NO DEPOSIT*NO RETURN." The bottom of the bottle is marked with the mold number (63-38B), 5 (place of manufacturing—plant 5), Anchor Hocking emblem (anchor over H), 49 (year of manufacturing—may also be 50 or 63), and the saying "Royal Ruby Anchorglass." There are actually three versions of this bottle: 63-38A, 63-38B, and 63-38C. One version is plain and two have quilted lines on the outside of the bottle. They were designed for foreign export but never produced.

Closeup of the neck and mouth of the Pfeiffer's Glasolite.

12 oz. beer bottle, 8 1/4", $80-100. The neck of the bottle is labeled "THROWAY PAY NO DEPOSIT" and "PFEIFFER'S GLASOLITE." The bottom of the bottle has the mold number (L-815), 2 (place of manufacturing—plant 2), Anchor Hocking emblem (anchor over H), and 47 (year of manufacturing). These bottles were made for another company two years before Royal Ruby beer bottles were sold by Schlitz. These bottles were never marketed.

Top left: 14 oz. beer bottle, 7", $25-40. The neck of the bottle is labeled "NOT TO BE REFILLED" and "NO DEPOSIT NO RETURN." The bottom of the bottle has the mold number (8548B) and three other numbers (6, 30, and 65). This was an experimental bottle never put into production.

Center left: Closeup of the neck and mouth of the experimental 14 oz. beer bottle.

Bottom left: 12 oz. Beer bottle, 6", $25-40. The neck of the bottle is labeled "DISPOSE OF PROPERLY" and "NO DEPOSIT*NO REFILL." Only the number 68 appears on the bottom of the bottle. This was another experimental bottle designed to have a "twist off" closure and 3/4" opening in the top of the bottle.

Bottom right: Closeup of the neck and mouth of the experimental 12 oz. beer bottle. Notice the 3/4" opening and the "twist off" closure design.

Bottles

12 oz. Beer bottle, 5 1/2",
$25-40. There are no
markings on this experimental bottle. It was designed to
have a "twist off" closure
and a 1" opening in the top
of the bottle.

Closeup of the neck and
mouth of the 12 oz.
experimental beer
bottle. Notice the 1"
opening and "twist off"
closure design.

Medicine Bottles

Aspirin bottle, 4 1/2", $20-25.
The bottom of the bottle has
the mold number (82-15A), 5
(place of manufacturing—
plant 5), Anchor Hocking
emblem (anchor over H), and
the saying "Royal Ruby
Anchorglass."

Medicine bottle, 4 3/4", $20-25.
The bottom of the bottle has the
mold number (50-14), 5 (place of
manufacturing—plant 5), and the
saying "Royal Ruby Anchorglass."

Chapter Three
Apothecary/Candy Jars

Anchor Hocking marketed six styles of candy and apothecary jars. At least the three pictured below were made in Royal Ruby. The 22 oz. stemmed candy jar was made in Forrest Green, but the color of the jar was "flashed" or applied as a coating. The "flashing" is not part of the glass and will flake off over time.

1971 catalog page listing descriptions of the apothecary/candy jars.

pictorial code	description	item number	BULK PACK dozen/carton	lbs/carton
A	22 oz. stemmed candy jar/cover	3291	1	11
B	12 oz. jar/cover	3249	1	8
C	10-oz. stemmed candy jar/cover	3290	1	7
D	24 oz. jar/cover	3285	1	11
E	12 oz. jar/cover	3284	2	15
F	35 oz. jar/cover	3286	1	13
G	2 lb. jar/cover	84	1	24
H	1/2 lb. jar/cover	82	2	17
I	4 oz. jar/cover	80	2	9
J	1 lb. jar/cover	83	1	15
K	4 lb. jar/cover	85	1/2	17
L	24 oz. stemmed Finlandia jar/cover	2624	1	15
M	10 oz. stemmed Finlandia jar/cover	2620	2	16
N	34 oz. stemmed Finlandia jar/cover	2634	1	17

NOTE: *See Index for additional listing of Finlandia items.*

1971 catalog page showing the six styles of apothecary/candy jars.

A   B   C   D   E   F

Left to right: 10 oz. stemmed candy jar with cover, 8 1/2", $10-15; 24 oz. candy jar with cover, 8 1/2", $15-20;
12 oz. candy jar with cover, 6 3/4", $10-15.

Chapter Four
Ash Trays

Anchor Hocking catalogs list over twenty-five types and styles of ash trays. They were made in crystal, avocado, honey gold, milk white, Forrest Green, and Royal Ruby. I have found the "anchor over H" emblem embedded in the glass on only some of the 5 3/4", 4 5/8", and 3 1/2" square ash trays. The mark is located in one corner of the base.

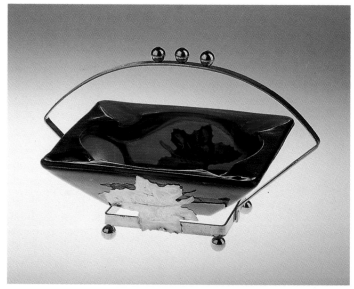

4 5/8" square ash tray in copper plated carrier, $10-12. Anchor Hocking did not make the carrier, but it is an interesting piece.

Left to right: 3 1/2" ash tray #R30, $5-8; 5 3/4" ash tray #R32, $10-15; 4 5/8" ash tray #R31, $8-10.

Four ash trays together so you can see the relative size of each. Clockwise from top left: 3 footed ash tray, 4", $5-8; square ash tray #R822, 4 1/4", $10-15; Queen Mary ash tray, 3 1/4", $5-8; grape leaf ash tray #R1274, 4 1/2", $5-7.

Queen Mary ash tray, 3 1/4", $5-8.

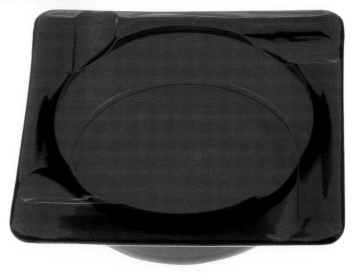

Square ash tray #R822, 4 1/4", $10-15.

Grape leaf ash tray #R1274, 4 1/2" in diameter, 7/8" high, $5-7. There is also a grape leaf 4 1/2" bowl. The handle of the bowl does not have the three grooves used to hold cigarettes and the bowl is 1 1/2" high.

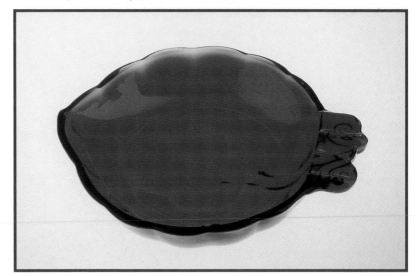

Early American Prescut ash tray, 7 3/4", $20-30.

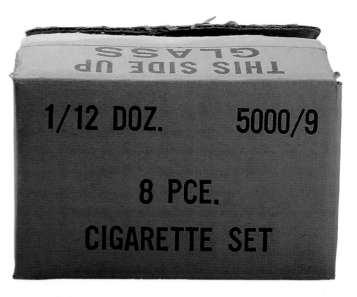

8-Piece Cigarette Set #5000/9, $100-125 complete, $15-20 for box only.

The 8-Piece Cigarette Set was called the "DEMON" Cigarette Set in the catalog. The set consists of one 4" x 6" cigarette box with Royal Ruby lid, $50-75, and six 3 1/2" square ash trays, $5-7 each.

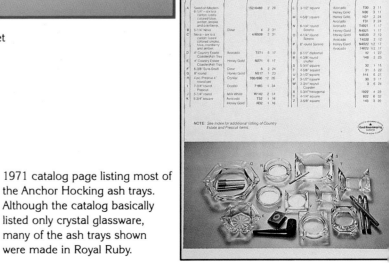

1971 catalog page listing most of the Anchor Hocking ash trays. Although the catalog basically listed only crystal glassware, many of the ash trays shown were made in Royal Ruby.

Chapter Five
Pitchers and Glasses

Many of the pitchers and glasses included in this chapter have also been listed under the specific pattern. This was done to make this book more "user friendly." I have also included many novelty tumblers, since they are interesting to collect and often provide dates useful in determining when certain patterns were produced. The 12 oz. "beer" tumbler was made by Anchor Hocking, but never listed in the catalogs. I have seven of these and all seven have the "anchor over H" emblem embedded in the bottom of the glass.

Baltic tumblers. Left to right: 10 oz. goblet #R3316, 4 1/2", $8-10; 5 oz. juice #R3311, 3 5/8", $5-8; 3 1/2 oz. cocktail #R3312, 2 5/8", $5-10.

Berwick stemware. Left to right: 9 oz. goblet #R736, 5 1/2", $15-20; 4 oz. juice/wine #R735, 4 1/2", $12-15; 6 oz. sherbet #R733, 3 1/2", $8-12; 3 1/2 oz. cocktail #R734, 3 7/8", $12-15.

Beverly 4 1/2 oz. cocktail
#R3264, 2 3/4" $10-12.

Early American. Left to right: 9 1/2 oz. goblet #R336, 5 1/4", $15-20; 5 1/2 oz. juice #R335, 4 1/4", $12-15;
6 oz. sherbet #R333, 4", $12-15; 4 1/2 oz. cocktail #R334, 3 3/8", $18-20.

Georgian. Left to right; 80 oz. pitcher, $40-50; 10 oz. pedestal tumbler, 5 1/2", $10-15; 14 oz. tumbler, 5 1/2", $10-15; 12 oz. tumbler, 5", $10-15; 9 oz. tumbler, 4 1/4", $10-15; 5 oz. tumbler, 3", $8-10; 5 oz. tumbler, 3 1/4", $8-10. It is difficult to find mint condition tumblers in this pattern. The projections approximately 2" below the rim tend to chip very easily. Some Georgian tumblers have the "anchor over H" emblem embedded in the glass.

High Point. Left to right: water pitcher #R1287, 2 1/2 qt., $75-100; 13 oz. ice tea #R1208, 5 1/2", $35-40; 9 oz. table tumbler #R1201, 4 1/4", $12-15; 5 oz. fruit juice #R1203, 3 5/8", $10-12. *13 oz. ice tea #R1208 courtesy of Rick Hirte.*

Hobnail. Left to right: pitcher #A2744, 60 oz., $30-35; 9 1/2 oz. tumbler #A2719, $8-10.

Old Café. Left to right: water tumbler, 4", $25-35; juice tumbler, 3", $20-25.

Pitchers and Glasses

Provincial. Left to right: ice lip pitcher #R1660, 64 oz., $50-60; 16 oz. ice tea #R1616, 5 3/4", $18-20; 12 oz. tumbler #R1612, 4 1/2", $12-15; 9 oz. old fashioned #R1609, 3 1/4", $12-15; 6 oz. fruit juice #R1606, 3 3/4", $8-12.

Roly Poly. Left to right: 96 oz. upright pitcher #R3687, $35-45; 13 oz. beverage/ice tea #R3658, 5", $8-10; 11 oz. tumbler, 4 3/4", $8-10; 9 oz. table tumbler #R3651, 4 1/4", $8-10; 5 oz. fruit juice #R3653, 3 3/8", $5-8. For additional pieces and information see Chapter 1.

Swirl. Left to right: 80 oz. jug #A1087, $35-45; 10 oz. tumbler #A1031, $8-10.

Windsor. Left to right: 60 oz. pitcher #A1153, $35-40; 9 oz. tumbler #A1131, 4", $8-10; 5 oz. fruit juice #A1133, 3 1/4", $8-10. I have only found the crystal tumblers with the "anchor over H" emblem embedded in the glass. For additional information see Chapter 1.

Whirly Twirly. Left to right: 3 quart water pitcher, $75-80; 12 oz. tumbler, 5", $15-20; 5 oz. tumbler, 3 1/2", $12-15. For additional pieces and information see Chapter 1.

Two tumblers often confused with the Whirly Twirly tumblers. Left to right: 9 1/2 oz. tall tumbler #R3597, 4 3/4", $10-12; 5 oz. juice, 3 3/4", $10-12. See Whirly Twirly in Chapter 1 for a comparison of these two tumblers with the Whirly Twirly tumblers.

Left to right: 16 oz. tumbler, 6", $10-15; 12 oz. tumbler, 5", $10-12; 6 1/2 oz. tumbler, 4", $8-10; 9 oz. tumbler, $8-10. This style is often referred to as the "Coupe" tumbler. This is not an established factory name. Some tumblers are marked with the "anchor over H" emblem embedded in the bottom of the glass.

Tapered tumblers. Left to right: 16 oz. tumbler, 5 3/4", $12-15; 12 oz. tumbler, 5 1/4", $10-15; 9 oz. tumbler, 4 1/4", $12-15.

"Hoe Down" Set of 12 oz. tapered tumblers. Left to right: "Do Si Do" tumbler, 5 1/4", $10-15; "Hoe Down" tumbler, 5 1/4", $10-15; "Swing her high, swing her low" tumbler, 5 1/4", $10-15; "Partners all" tumbler, 5 1/4", $10-15.

The "Hoe Down" Set of tapered tumblers was produced in two sizes, the 16 oz. tumbler on the left and the 12 oz. tumbler on the right.

Royal Ruby

Tapered tumblers. Left to right: 12 oz. tumbler with lily of the valley flowers, 5 1/4", $10-15; 12 oz. grape tumbler, 5 1/4", $10-15; 16 oz. tumbler with gold leaves, 5 3/4", $12-15.

8 Piece set of 12 oz. grape tumblers #R400/75-GRAPE, 5 1/4", $60-75.

Fluted tumblers. Left to right: 13 oz. tumbler, 5 5/8", $12-15; 11 oz. tumbler, 5 1/2", $10-15; 9 1/2 oz. tumbler, 5 1/4", $10-15; 5 oz. tumbler, 3 3/4", $12-15.

Coupe novelty tumblers. Left to right: 16 oz. "Sun Valley" tumbler, 6", $10-12; 16 oz. "USS Alabama" tumbler, 6", $10-12; 6 1/2 oz. "Silver Dollar City" tumbler, 4", $5-8.

Coupe novelty tumblers. Left to right: 12 oz. "Mackinac Bridge" tumbler, 5", $8-10; 12 oz. "San Francisco" tumbler, 5", $8-10; 12 oz. "Akdar Temple 1978" tumbler, 5", $8-10.

Coupe novelty tumblers. Left to right: 12 oz. "Pioneer Women, Ponca City, Oklahoma" tumbler, 5", $8-10; 12 oz. "Hershey Park, Hershey PA" tumbler, 5", $8-10; 12 oz. "Arizona" tumbler, 5", $8-10.

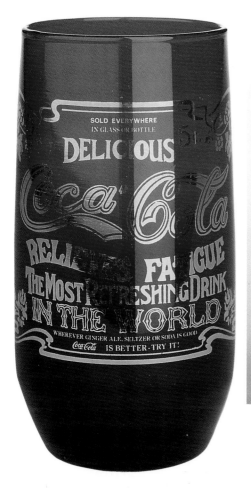

Roly Poly novelty tumblers. Left to right: 13 oz. "The old Union Depot, Kansas City 1850-1950" tumbler, 5", $8-10; 13 oz. "Nelson Art Galleries, Kansas City 1850-1950" tumbler, 5", $8-10; 13 oz. "1950 Seventy Third and First Consolidated Convention, American Flint Glass Workers Union, Toledo, Ohio" tumbler, 5", $8-10.

Absolutely gorgeous 12 oz. "Coca Cola" tumbler, 5", $25-30. The inscription reads, "Sold everywhere in glass and bottle. Delicious Coca Cola relieves fatigue. The most refreshing drink in the world. Wherever ginger ale, seltzer, or soda is good, Coca Cola is better—try it!"

Coupe novelty tumblers. Left to right: 9 oz. "Arizona" tumbler, 3", $5-8; 9 oz. "Central City, Colorado" tumbler, 3", $5-8; 9 oz. "Chicago" tumbler, 3", $5-8; 9 oz. "Dallas Texas 1978" tumbler, 3", $5-8.

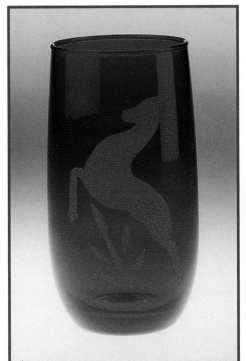

13 oz. Roly Poly tumbler with
etched deer, 5", $8-10.

9 oz. Roly Poly tumbler used to market
cottage cheese, 4 1/4", $15-20. The
inscription on the lid states, "Farm
Maid Creamed Cottage Cheese 10 oz.
net".

10 oz. Baltic "Howdy Wrangler" tumbler,
4 1/2", $12-15. I have only found these
tumblers in El Reno, Oklahoma, and St.
Louis, Missouri.

Left to right: 80 oz. ice lip pitcher, $75-100; 20 oz. tumbler, 6", $15-20; 12 oz. tumbler, 4 3/4", $12-15. Most of the 12 oz. tumblers I have found have the "anchor over H" emblem embedded in the bottom of the glass.

Left to right: 9 1/2 oz. ship design tumbler #A3519/712, 4 3/4", $10-12; 9 1/2 oz. small flower design tumbler #A3519/714, 4 3/4", $10-12; 9 1/2 oz. large flower design tumbler, 4 3/4", $10-12; 9 1/2 oz. wild geese design tumbler #A3519/713, 4 3/4", $10-12; 9 1/2 oz. laurel wreath design tumbler #A3519/711, 4 3/4", $10-12.

16 oz. long boy ice tea, 6 1/2", $12-15.

Left to right: 9 1/2 oz. top gold band and hairline tumbler, 4 3/4", $10-12; 9 1/2 oz. gold band and hairline tumbler #A3519/11, 4 3/4", $10-12; 9 1/2 oz. three gold line tumbler, 4 3/4", $10-12. It is difficult to find these tumblers with the gold lines in good shape. Price stickers that are not water soluble, if placed over the lines, will pull the lines off when the sticker is removed. Also, cleaning the glass with anything abrasive will remove the delicate lines.

9 1/2 oz. tall tumbler #R3321, 4 5/8", $12-15. The catalog also lists a 5 oz. fruit juice #R3323 in Royal Ruby. There is a 12 oz., 5" Forrest Green tumbler that may also exist in Royal Ruby.

12 oz. ice tea glass, 4 7/8", $12-15.

Left to right: 9 oz. balled stem goblet, 5 1/4",
$10-15; 2 1/2 oz. footed wine #R1755, 3 3/8",
$12-15.

12 oz. "beer" tumbler, 5 1/2", $12-15. This is
an extremely heavy glass with the "anchor over
H" emblem embedded in the bottom of the
glass.

3 1/2 oz. footed cocktail #R3662, 3", $15-20.

Chapter Six
Relish Trays/Lazy Susans

The majority of the relish trays/lazy susans were not sold by Anchor Hocking. They were included in the book because the trays used Old Café inserts. With the exception of the ornate silver set, the crystal, Forrest Green, and Royal Ruby inserts are interchangeable in the trays.

The ornate silver relish set requires the unlined inserts so you can see the engraving in the bottom of the tray. While Royal Ruby Old Café glassware is not extremely abundant, the relish tray inserts must have been made in great quantity.

Complete Old Café relish set, $125-150, consists of one 15" crystal plate, $30-35, five inserts, $8-10, one 7" covered dish, $10-15, and one 4" two-piece rotating base, $30-40.

Anchor Hocking made four different relish tray inserts: Forrest Green, $8-10; Royal Ruby, $8-10; crystal with lines on the bottom, $5-8; crystal with no lines on the bottom, $5-8.

Crystal plate, 15", $30-35. There is a ridge around the clear area in the center of the plate. This ridge holds the base in place so the tray doesn't move off center and tip over.

Crystal two-piece rotating base for the relish set, $30-40.

Royal Ruby inserts and crystal covered dish in a stainless steel tray with brass handles (maker unknown). This set does not have a base. The complete set, $75-100.

Art deco relish set but different in color only (maker unknown). The complete set, $50-75.

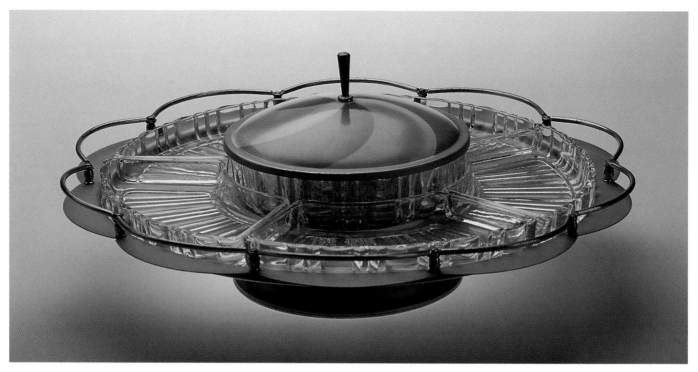

Art deco relish set (maker unknown). The set rotates on a metal base attached to the tray. The railing around the tray is brass. The complete set, $50-75.

Overhead view of same relish set.

Gold and white art deco relish set (maker unknown). This set also rotates on a metal base attached to the tray. The complete set, $50-75.

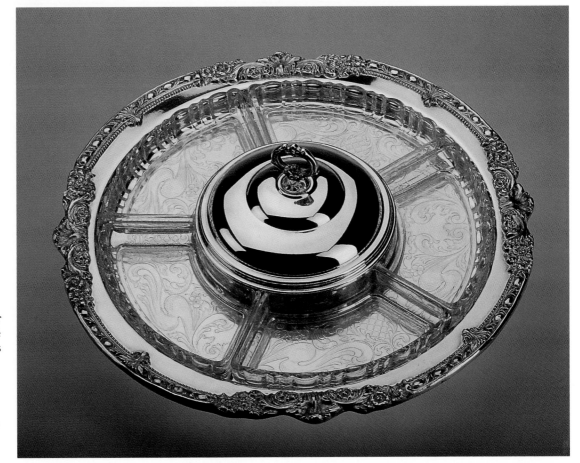

Ornate silver relish set made by Sheridan Silver Company. The Old Café inserts do not have lines in the bottom so the relish tray engraving is visible. The set rotates on an attached metal base. The complete set, $125-150.

Chapter Seven
Serva-Snack/Popcorn/Punch Sets

Many of the items made by Anchor Hocking in Royal Ruby were sold in packaged sets. The three serva-snack sets shown are the only three that were sold with Royal Ruby 5 oz. cups. The popcorn set can be purchased cheaper as a set than the pieces separately. These sets were often stored for years and subjected to moisture. This would account for the all too common water stains and poor condition of many of the boxes.

14-Piece Punch Set consists of one 10" bowl, one bowl base, and 12 punch cups, $100-125 for complete set, $20-25 for box only.

10" Punch bowl, $30-40; bowl base, $35-50.

Punch bowl base, $35-50.

Punch cup, $2-4.

9-Piece Popcorn Set #R200/55, consists of one popcorn bowl (punch bowl without the base) and eight 5 1/4" bowls, $12-15. The complete set $90-100, for box only $20-30.

The popcorn bowl may be confused with the Baltic 12 oz. footed bowl #R3322. The 5 1/4" popcorn bowl on the left has considerably thicker glass than the Baltic 12 oz. footed bowl. See Baltic for a comparison of the two bowl bases.

8-Piece Serva-Snack Set #200/27 with four Royal Ruby cups and four crystal trays, $25-35.

8-Piece Serva-Snack Set #R200/42 with four Royal Ruby cups and four crystal trays, $25-35.

Serva-Snack tray, $4-5; 5 oz. cup, $2-4.

8-Piece Serva-Snack Set #R200/42 in another style of box, $25-35.

Serva-Snack tray, $4-5; 5 oz. cup, $2-4.

Fan shaped serva-snack crystal tray, $4-5; 5 oz. cup, $2-4.

Chapter Eight
Vases

Over the years, Anchor Hocking produced a variety of vases. Most of the vases were trademarked with paper labels. The only vases I have found trademarked with the "anchor over H" emblem embedded in the glass are the crimped top vase #R3306, crimped bud vase #R3303, and the Rainflower vase. The trademark on the Rainflower vase is extremely hard to see, but it can be found by using a magnifying glass.

Many of the vases were sold to other companies where they were etched with flowers or figures. The etching process required the piece to be first coated with paraffin wax. The design was then scraped into the wax and the piece subjected to a hydrofluoric acid mist or vapor. The acid dissolved the glass in areas not protected by the wax. After the piece was washed to remove the acid, the paraffin wax was melted off. This left the design, in white, on the surface of the glass.

The "Wilson" 4" ivy ball #R3354 and "Coolidge" 6 3/8" vase #R3346 were used to market mosquito repellent candles. Although only two candles are pictured, there are undoubtedly numerous other companies that made the candles. The "Wilson" 4" ivy ball #R3354 can be found in a variety of wall hangers. I have seen many different styles over the years, but only included a couple in the book.

Left to right: "Harding" vase #R3345, 6 3/8", $8-10; crimped top vase #R3306, 6 1/2", $8-10; crimped bud vase #R3303, 5 3/4", $8-10. Both the crimped top vase #R3306 and #R3303 have the "anchor over H" embedded in the glass on some vases.

Left to right: bud vase #R3302, 3 3/4", $5-8; crimped top vase, 4 1/4", $8-10; bud vase #R3301, 4", $8-10.

Left to right: ivy ball, 6", $12-15; ivy ball, 4", $8-10. Notice both ivy balls have scalloped openings.

Left to right: top of "Wilson" 4" ivy ball #R3354 (not scalloped) and the scalloped 4" ivy ball.

The "Wilson" 4" ivy balls were extremely popular and sold to other companies. The ivy ball on the left was given a metal ring and chains so it could be hung. The center ivy ball was sold by Anchor Hocking. The ivy ball on the right was marketed as an insect repellent candle.

Vases

The top reads: "Moskeeto Lites, Victrylite Candle Co. Oshkosh Wis, Net weight min. 10 oz. Active ingredients per weight 7% 2-phenyl cyclohexanol, 9% 2-cyclohexyl cyclohexanol, vapor booster saturated with 1.5 grams cyclohexyl cyclohexanol, 3.5 grams 2-phenyl cyclohexanol, inert ingredients 90% scented paraffin waxes, direction for use on the box."

Another "Wilson" ivy ball sold in a copper plated metal holder. The ivy ball can swing back and forth in the cradle. Maker of holder unknown, $10-15.

"Wilson" ivy ball mounted in a metal bracket for hanging on the wall. Maker of bracket unknown, $10-15.

Closeup of insect repellent candle. The label states "Insecto Candle, ingredients: 5% dimethyl phthlate, 1% perfume oil, inert 94%, Interstate Chemical Company, St. Joseph, MO."

The "Coolidge" 6 3/8" vase #R3346 was another Anchor Hocking vase marketed by other companies. Left to right: "Coolidge" vase sold by Anchor Hocking, $8-10; insect repellent candle sold by the Interstate Chemical Company, $15-20; vase distributed to members of the Masons, $8-10.

The directions for, opposite page bottom right photo, use of the Insecto Candle were attached to the bottom of the base. The directions read, "Insecto Candles, Directions: Light candle, place nearby. Effective approx. in 20 sq. ft. area. Net 8 oz. Ingredients: 5% Dimethyl Phthlate, 1% perfume oil, Inert, 94%. Interstate Chemical Co., St. Joseph, Mo."

"Coolidge" vase showing the etched Mason symbol.

The "Coolidge" vase was also engraved and sold as novelties. This vase is engraved with "Mary Brier, New Orleans, 1954." In the absence of Anchor Hocking records, these novelties are useful in determining when items were manufactured, $5-10.

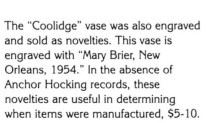

There are two versions of the "Hoover" 9" vase #R53. On the left is the plain version, $15-20. The vase on the right has the image of two birds feeding four chicks in a nest located in the branches of a flowering tree, $15-25.

There are two versions of the 9" vase #R597. On the left is the plain rimmed version produced in the late 1940s, $30-35. On the right is the crimped top version produced in the 1950s, $30-35.

Rainflower vase, 5 1/2", $10-15. The Rainflower vases are marked with the "anchor over H" emblem embedded in the glass. Because the bottom of the vase has many imperfections, the mark is hard to see without a magnifying glass, but it is there.

Crimped top vase, 6 7/8", $12-15. This gorgeous vase was originally sold to florist shops. Unlike other vases produced by Anchor Hocking, the outer surface of the vase is smooth and the swirled lines are applied to the inner surface.

Vase, 4", $10-15. Without an established catalog reference or name, I have seen this vase called a "female spittoon."

Bud vase, 7 3/4", $20-30.

Crimped top vase #R3308, 7", $20-30. This vase is easily found in green, but rarely found in Royal Ruby.

Chapter Nine
Miscellaneous/Novelty Items

    Throughout the period of 1940 to 1964, Anchor Hocking made countless novelties in Royal Ruby. Very few were trademarked in the glass. The majority of the pieces were marked with paper labels. Also included in this chapter are items which didn't seem to fit in any other chapter.

Scalloped swirled bowl, 6 3/8",
$12-15. The swirls in the bowl are
found on the inner surface, the
outer surface is smooth.

Maple leaf bowl, 6 5/8", $8-10.

Opposite page:
Top: Left to right: dessert bowl, 4 1/2", $5-10; berry bowl, 8 1/2",
$15-20.

Bottom: Oval vegetable bowl, 8 1/2" long, $15-20. This bowl is
extremely common in the central United States.

Handled mint tray #A936, 7 1/4",
$10-15.

Three-toed nappy #A588 with handle, 7", $25-30 with handle, $15-20 without handle.

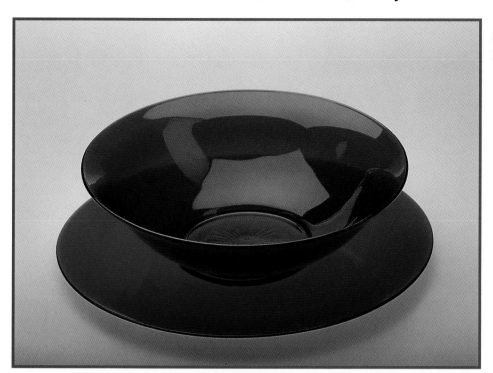

Salad bowl, 11 1/2", $30-40, underliner plate, 13 1/2", $30-40.

Left to right: dessert bowl #R1074, 4 1/2", $8-10; large bowl #R1078, 8", $15-20.

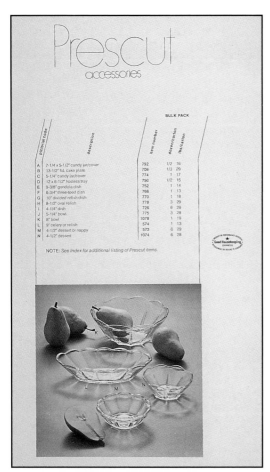

Prescut accessories

pictorial code	description	item number	BULK PACK dozen/carton	lbs/carton
A	7-1/4 x 5-1/2" candy jar/cover	792	1/2	16
B	13-1/2" ftd. cake plate	706	1/3	29
C	5-1/4" candy jar/cover	774	1	17
D	12 x 6-1/2" hostess tray	750	1/2	15
E	9-3/8" gondola dish	752	1	14
F	6-3/4" three-toed dish	768	1	13
G	10" divided relish dish	770	1	18
H	8-1/2" oval relish	778	3	29
I	4-1/4" dish	726	6	26
J	4-1/4" bowl	775	3	28
K	8" bowl	1078	1	19
L	9" celery or relish	574	1	13
M	4-1/2" dessert or nappy	573	6	29
N	4-1/2" dessert	1074	6	28

NOTE: See Index for additional listing of Prescut items.

Bon Bon #A573, 6 1/2", $10-15. You will notice the factory bulletin shows three rows of bumps around the base of the bowl, however, the bowl pictured here does not have the bumps. There may be two versions of the bowl.

Grape leaf bowl, 4 1/2", $8-10. This bowl has the same shape as the grape leaf ash tray except the handle does not have the three grooves to rest cigarettes in and the bowl is taller than the ash tray.

Relish dish #A560, 7 3/8" x 3 3/4", $8-12. There are two types of this relish dish. They have different patterns in the bottom, either straight lines or quilts. The tray pictured here has the straight line pattern while the tray listed in the factory bulletin is quilted.

Opposite page:
Top left: 1971 catalog page showing both the 1074 and 1078 bowls in crystal. Notice there are two other bowls in this design that may also exist in Royal Ruby.

Bottom: Console Set, $50-75. This set uses two R1074 and one R1078 bowls. The bowls were sold to another company for conversion into the console set.

Top right: Closeup of candle holder showing the shape of the stem. This is the only stem pattern I have been able to find.

Relish dish, 4 1/2" x 6", $10-15.

Ice bucket with tongs, $50-75 with tongs, $5-10 for tongs.

Left to right: 15 hr. candle tumbler #R373, 2 1/2" high, $3-5; 10 hr. candle tumbler #R824, 2" high, $2-4. Many of the candle tumblers are marked with the "anchor over H" emblem embedded in the glass.

Left to right: small cocktail shaker, 5 1/2" to top of chromed lid, $25-35; large cocktail shaker, 9 1/2" to top of chromed lid, $75-100.

Crystal puff box with Royal Ruby cover #E522, 4 5/8", $15-20 complete.

Crystal cigarette or jewel box with Royal Ruby cover #E599, 4 1/4", $20-25 complete.

Crystal marmalade jar with Royal Ruby cover #E514, 3 5/8" x 5 1/8", $15-25 (complete set). You will notice a 6" diameter under liner plate beneath the marmalade jar. This was sold as an accessory item. Today many people mistakenly sell the under liner plates as saucers. Under liner plate $5-10; Royal Ruby lid only $8-10.

Crystal marmalade with Royal Ruby cover, $15-25. This version uses the two-handled crystal Early American Prescut bowl.

Clockwise from top left: marmalade under liner plate with a 2 1/2" diameter depression; R-1700 saucer with a 2" diameter depression; R-4000 saucer with a 1 7/8" diameter depression.

Crystal candy jar with Royal Ruby cover #E775, 5 1/2",
$15-20.

Factory sheet announcing
new accessory pieces of
crystal and Royal Ruby.

Left to right: plain stemmed sherbet, $5-10; ball stemmed sherbet, $5-10; Baltic 6 1/2 oz. sherbet #R3313, $5-10.

Novelty item that resembles the 4" ivy ball with a crystal lid, 5 1/2" total height, $10-15. The lower Royal Ruby ball is marked with the "anchor over H" emblem embedded in the glass.

"Sunburst" relish dish, 5 3/4" x
7 1/2", $15-20. This dish is
also available in crystal.

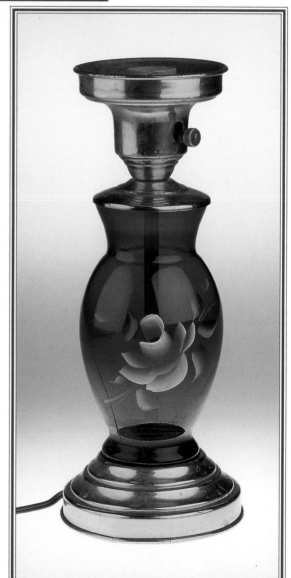

Lamp using the "Coolidge" 6 3/8"
vase, $50-75. The vases were sold
to another company for conversion
into lamps. The lamps were also
made with the "Hoover" 9" vase.

Chapter Ten
Sugars and Creamers

There can be some confusion about the three styles of sugars and creamers produced by Anchor Hocking, especially between the R-1700 line produced in the 1940s and the R-4000 line produced in 1956-1957. This is why I have included detailed photographs of the handles. I have not found any of the three styles of sugars and creamers with the trademark embedded in the glass.

Left to right: sugar without lid, $8-10; creamer $8-10.

Left to right: R-4000 sugar, $8-10; creamer, $8-10.

Left to right: R-1700 flat sugar, $8-10; creamer, $8-10.

Closeup of R-4000 sugar and creamer handle.

Closeup of sugar and creamer handle.

Closeup of R-1700 sugar
and creamer handle.

Chapter Eleven
Factory Sheets

Information about the production of Anchor Hocking glassware is limited. I was able to obtain certain "jobber" sheets that were distributed in place of catalogs. Anchor Hocking relied on these sheets from 1945 to the middle 1950s. After that, catalogs were published and distributed.

Provincial Pattern.

NEW *Royal Ruby* STEMWARE
WITH CRYSTAL STEMS

Anchor Hocking again leads in the creation of a colorful, appealing line of stemware in its new Royal Ruby line. The rich red bowl and the contrasting crystal stem combine to make a stemware service of pleasing acceptance to any home-maker. The beauty of this ware makes it of particular interest for gifts for all occasions. Like all other Anchor Hocking products it is priced low to make it doubly attractive.

Berwick pattern

R336—10 OZ. GOBLET Packed 3 doz. carton, 16 lbs.	**R335—4½ OZ. FRUIT JUICE** Packed 3 doz. carton, 10 lbs.	**R334—3½ OZ. COCKTAIL** Packed 3 doz. carton, 10 lbs.	**R333—6 OZ. SHERBET** Packed 3 doz. carton, 16 lbs.

SET PACKING

R300/144—8 PCE. **GOBLET SET** Set of 8 in Gift Carton, 6 Sets to a Shipper, 27 lbs. COMPOSITION: 8 only R336 Goblets	**R300/145—8 PCE.** **FRUIT JUICE SET** Set of 8 in Gift Carton, 6 Sets to a Shipper, 16 lbs. COMPOSITION: 8 only R335 Fruit Juice Glasses	**R300/146—8 PCE.** **COCKTAIL SET** Set of 8 in Gift Carton, 6 Sets to a Shipper, 14 lbs. COMPOSITION: 8 only R334 Cocktails	**R300/147—8 PCE.** **SHERBET SET** Set of 8 in Gift Carton, 6 Sets to a Shipper, 24 lbs. COMPOSITION: 8 only R333 Sherbets

A spotlighted Feature Display of Royal Ruby Stemware will bring excellent sales results. We recommend an adequate under-stock when you put it on sale. It will sell in big volume and that is the kind of merchandise you need today.

READY FOR IMMEDIATE SHIPMENT.

ANCHOR HOCKING GLASS CORPORATION, LANCASTER, OHIO, U. S. A.

According to the 1971 catalog, this is not the Berwick pattern. Someone wrote the incorrect pattern on the sheet.

Royal Ruby Anchorglass

R3653—5 oz. Fruit Juice
Pkd. 6 doz. ctn.—wt. 18 lbs.

R3651—9 oz. Table Tumbler
Pkd. 6 doz. ctn.—wt. 27 lbs.

R3658—13 oz. Iced Tea
Pkd. 6 doz. ctn.—wt. 33 lbs.

R3687—3 Qt. Water Pitcher
Pkd. 1 doz. ctn.—wt. 37 lbs.

R3312—3½ oz. Ftd. Cocktail
Pkd. 6 doz. ctn.—wt. 19 lbs.

R3311—5 oz. Ftd. Fruit Juice
Pkd. 6 doz. ctn.—wt. 22 lbs.

R3316—10 oz. Ftd. Goblet
Pkd. 6 doz. ctn.—wt. 37 lbs.

R3313—6½ oz. Ftd. Sherbet
Pkd. 6 doz. ctn.—wt. 27 lbs.

R828—6¼" Sherbet Plate
Pkd. 6 doz. ctn.—wt. 35 lbs.

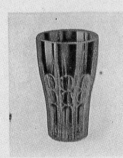

R1203—5 oz. Fruit Juice
Pkd. 6 doz. ctn.—wt. 23 lbs.

R1201—9 oz. Table Tumbler
Pkd. 6 doz. ctn.—wt. 33 lbs.

R1208—13 oz. Iced Tea
Pkd. 6 doz. ctn.—wt. 52 lbs.

R1287—2½ Qt. Water Pitcher
Pkd. 1 doz. ctn.—wt. 37 lbs.

R3662—3½ oz. Ftd.
Cocktail
Pkd. 6 doz. ctn.—
wt. 23 lbs.

R3323—5 oz. Fruit Juice
Pkd. 6 doz. ctn.—
wt. 21 lbs.

R3321—9½ oz. Tall
Tumbler
Pkd. 6 doz. ctn.—
wt. 27 lbs.

R1755—2½ oz. Ftd. Wine
Pkd. 6 doz. ctn.—
wt. 15 lbs.

*Trade Mark

ANCHOR HOCKING GLASS CORPORATION, LANCASTER, OHIO, U. S. A.

NEW RUBY TUMBLERS
AND WATER PITCHERS

HOBNAIL PATTERN
(BLOWN)

A2744—60 OZ.
PITCHER
Pkd. 1 doz. ctn.—
wt. 28 lbs.

A2719—9½ OZ.
TUMBLER
Pkd. 12 doz. ctn.
—wt. 48 lbs.

ANCHOR HOCKING GLASS CORPORATION
LANCASTER, OHIO, U. S. A.

PRINTED IN U.S.A.

★ R3653

★ R3651

★ R3658

★ R3687

★ R3316

R2279 — R2229 — R2275 — R2241
★ R2200/4

DESCRIPTION

ROYAL RUBY
Anchorglass*

ROLY-POLY LINE

★ **R3653—5 OZ. FRUIT JUICE**
Pkd. 6 doz.—wt. 19 lbs.

★ **R3651—9 OZ. TABLE TUMBLER**
Pkd. 6 doz.—wt. 26 lbs.

★ **R3658—13 OZ. ICED TEA**
Pkd. 6 doz.—wt. 35 lbs.

★ **R3687—3 QT. WATER PITCHER**
Pkd. 1 doz.—wt. 40 lbs.

SETS

★**R3600/2—9 PCE. ICED TEA SET**
Each Set Pkd. in Ind. C/D Ctn.—
6 in Outer R/S Ctn.—wt. 50 lbs.
COMPOSITION:
Eight R3658 Iced Teas
One R3687 Pitcher

★**R3600/4—24 PCE. REFRESHMENT SET**
Each Set Pkd. in Ind. R/S Ctn.
—wt. 9 lbs.
COMPOSITION:
Eight R3653 Fruit Juices
Eight R3651 Table Tumblers
Eight R3658 Iced Teas

R3600/5—19 PCE. REFRESHMENT SET
Each Set Pkd. in Ind. R/S Ctn.
—wt. 11 lbs.
COMPOSITION:
Six R3653 Fruit Juices
Six R3651 Tumblers
Six R3658 Iced Teas
One R3687 Pitcher

FOOTED GOBLET

★**R3316—10 OZ. FTD. GOBLET**
Pkd. 6 doz.—wt. 38 lbs.

"CHARM" TABLEWARE

R2279—CUP
Pkd. 6 doz.—wt. 28 lbs.

R2229—5⅜" SAUCER
Pkd. 6 doz.—wt. 31 lbs.

R2275—4¾" DESSERT
Pkd. 6 doz.—wt. 37 lbs.

R2241—8⅜" PLATE
Pkd. 3 doz.—wt. 40 lbs.

16 PCE. LUNCHEON SET

★**R2200/4—16 PCE. LUNCHEON SET**
Each Set Pkd. in Ind. R/S Ctn.
—wt. 10 lbs.
COMPOSITION:
Four R2279 Cup
Four R2229 Saucer
Four R2275 Dessert
Four R2241 Plate

*Reg. U. S. Pat. Off.

10

THE "WINDSOR" PATTERN

A Table Tumbler and Fruit Juice Glass in a design of sparkling brilliance—the "Windsor" with Pitcher to match.

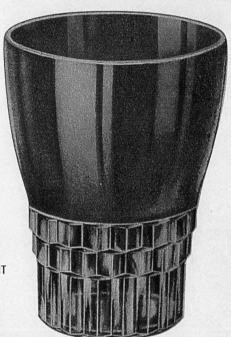

**LIGHT WEIGHT
PRESSED**

A1131—9 OZ. TUMBLER—
Ruby
Pkd. 12 doz. ctn.—wt. 65 lbs.

A1133—5 OZ. FRUIT JUICE—
Ruby
Pkd. 12 doz. ctn.—wt. 45 lbs.

**DISPLAY SETS AND
PRICE BOTH SETS
AND INDIVIDUAL
PIECES**

Blown Pitcher

A1153—60 OZ. PITCHER—
Ruby
Pkd. 1 doz. ctn.—wt. 28 lbs.

HOW TO DISPLAY RUBY

To properly display Ruby, use a white surface and background, and stand each Tumbler upright. Do not nest them.

Royal Ruby Anchorglass *

R1779—Cup
Pkd. 6 doz. ctn.—wt. 28 lbs.

R1729—Saucer
Pkd. 6 doz. ctn.—wt. 27 lbs.

R1738—7¾" Salad Plate
Pkd. 4 doz. ctn.—wt. 32 lbs.

R1741—9⅛" Dinner Plate
Pkd. 4 doz. ctn.—wt. 44 lbs.

R1774—4¼" Fruit
Pkd. 12 doz. ctn.—wt. 48 lbs.

R1778—8½" Bowl
Pkd. 2 doz. ctn.—wt. 34 lbs.

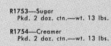

R1767—7½" Soup Plate
Pkd. 4 doz. ctn.—wt. 35 lbs.

R1753—Sugar
Pkd. 2 doz. ctn.—wt. 13 lbs.

R1754—Creamer
Pkd. 2 doz. ctn.—wt. 13 lbs.

R3345—6⅜" Vase
Pkd. 4 doz. ctn.—wt. 28 lbs.

R3346—6⅜" Vase
Pkd. 4 doz. ctn.—wt. 28 lbs.

R597—9" Vase
Pkd. 1 doz. ctn.—wt. 30 lbs.

R53—9" Vase
Pkd. 2 doz. ctn.—wt. 38 lbs.

R3354—4" Ivy Ball
Pkd. 6 doz. ctn.—wt. 27 lbs.

R1274—4½" Ash Tray
Pkd. 6 doz. ctn.—wt. 26 lbs.

*Trade Mark

ANCHOR HOCKING GLASS CORPORATION, LANCASTER, OHIO, U. S. A.

(SEE OTHER SIDE)

Anchorglass

To highlight any room ROYAL RUBY Glassware

New and exciting accent pieces in lovely radiant ruby including dainty Bud Vases, sparkling Dessert Dishes, smart Swedish Ash Trays and charming Apothecary and Candy Jars. They'll appeal to decor minded customers—and everyone who likes nice things. They're perfect as collector's items which will increase in value through the years. Every item will be a seller!

No.	Size	Item	Doz. Ctn.	Lbs. Ctn.
R3303	5¾"	Bud Vase	4	22
R3301	4"	Bud Vase	4	22
R824*	10 Hr.	Candle Tumbler	6	15
R373*	15 Hr.	Candle Tumbler	6	17
R1078	8"	Large Bowl	1	17
R1074	4½"	Dessert	6	29
R32	5¾"	Ash Tray	1	15
R31	4⅝"	Ash Tray	3	22
R30	3½"	Ash Tray	3	11
R3285	24 oz.	Apothecary Jar & Cover	1	12
R3290	10 oz.	Stemmed Candy Jar & Cov.	1	7

*1 Dozen in Chip Box — 6 to shipping carton

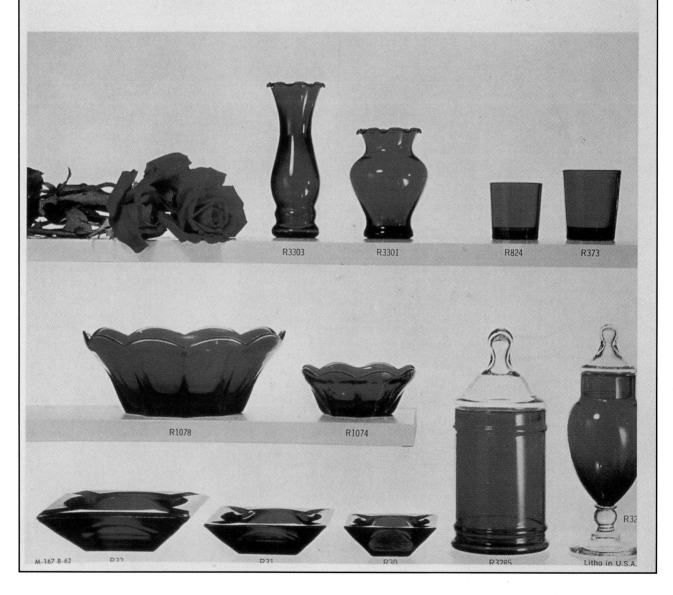

R3303   R3301   R824   R373

R1078   R1074

M-167-8-62   R32   R31   R30   R3285   Litho in U.S.A.

DISTINCTIVE

Royal Ruby

FOR HOTELS, RESTAURANTS and CLUBS!

NO.	SIZE	ITEM	DOZ. CTN.	LBS. CTN.
R49	9 oz.	Georgian Tumbler	1	31 *
R3555	10½ oz.	Footed Goblet	3	17
R30	3½"	Sq. Ash Tray	3	11
R31	4⅝"	Sq. Ash Tray	3	22
R32	5¼"	Sq. Ash Tray	1	15
R3303	5¼"	Crimped Bud Vase	4	22

* 3 Cartons to Outer

another fine
ANCHOR HOCKING
product

LANCASTER, OHIO.

R49

R3555

DISCONTINUED

R30

R31

R32

R3303

HR-22
8-62

LITHO IN U.S.A.

Notice the 10 1/2 oz. footed goblet #R3555 tumbler was marked "DISCONTINUED." I could not find out if the goblet was discontinued after some goblets were sold, or before production began. I have never seen this piece for sale.

Occasional Pieces in
Ruby Glass

A875—5¼" NUT DISH—Ruby
Packs 2 doz. ctn.—18 lbs.

A573—6½" BON BON—Ruby
Packs 2 doz. ctn.—16 lbs.

A560—7⅜"x3¾" RELISH DISH—
Ruby
Packs 2 doz. ctn.—19 lbs.

A977—8" FOOTED TRAY—Ruby
Packs 2 doz. ctn.—23 lbs.

A936—7¼" HANDLED MINT
TRAY—Ruby
Packs 2 doz. ctn.—18 lbs.

A588—7" THREE-TOED NAPPY
—Ruby
Packs 2 doz. ctn.—20 lbs.

A597—8¾" VASE—Ruby
Packs 1 doz. ctn.—27 lbs.

ANCHOR HOCKING GLASS CORPORATION
HOCKING DIVISION
LANCASTER, OHIO, U. S. A.

PRINTED IN U.S.A.